Contents

T0268760

Introduction   iv

Biology

Chapter 1   The characteristics of living things   1
Chapter 2   Identifying species   8
Chapter 3   Cells   13
Chapter 4   Microorganisms   22

Chemistry

Chapter 5   The states of matter   29
Chapter 6   Atoms and elements   33
Chapter 7   Elements, compounds and mixtures   38
Chapter 8   Physical properties of matter   45
Chapter 9   Chemical reactions   52
Chapter 10   Acids and alkalis   55

Physics

Chapter 11   Measurement   59
Chapter 12   Energy   66
Chapter 13   Sound   72
Chapter 14   Electricity   76

Earth and space

Chapter 15   The Earth in space   82
Chapter 16   A closer look at the Earth   87

Introduction

The aim of every science course is to help you become scientifically literate or, more simply, to help you become a 'scientific citizen'. This means that you can confidently talk and write about the science you have studied and know how it helps us to understand and live in our world. Below are some questions from this book that a scientific citizen should be able to answer. Just read through them slowly.

How can you tell if something is alive? How are plants different from animals? What do you find inside a cell? Name three kingdoms of microorganisms. What is the link between elements and atoms? What does the pH scale measure? Name two places where energy is stored. What does a vibrating object produce in the air around it? What does an ammeter measure? How did the solar system form and what is it like inside the Earth? How are scientific enquiries carried out and how can you find out information from them?

The chances are that you will not be able to answer most of those questions now, but if you work through this book as you complete the chapters in the Checkpoint Science Stage 7 Student's Book, you will be on your way to being scientifically literate.

Here is the first challenge of the book – look at these questions again and write down any answers you might have. It does not matter if you cannot think of any answers to a question, just keep a record of your answers for later. When you have completed this workbook, look back at these questions again, write down your answers and see how they differ from your first answers. This should show you that you are well on your way to being a scientific citizen.

Most of the questions aim to test your knowledge and understanding of science, but some questions have this icon ⭐. These questions aim to test your science enquiry skills.

Some other questions have this icon 🔗. These are questions about using models to help you learn about and understand scientific ideas.

Yet other questions have this icon 🧩. These questions put science in context.

So now it is time to start. Read each question carefully, think about it, then write down the answer in the space provided, or as otherwise instructed. Often you will have to write down some facts or explanations, sometimes you will need to tick a box and occasionally you will have to link details by lines or construct and interpret graphs.

1 The characteristics of living things

Living and never lived

⭐ **1** Is the hypothesis 'A plant can show it is alive by growing' testable? Explain your answer.

....

....

Looking for characteristics of life

2 A rabbit is eating grass. Its ears turn to face a sound and the rabbit stops eating. As the sound gets louder the rabbit hops away.

a How many characteristics of life is the rabbit displaying?

....

b Name each characteristic of life and copy down the text that describes it from the question paragraph.

....

....

....

Nutrition

⭐ **3** Su Lin is looking at animals with pointed teeth and flat-topped teeth. She constructs the hypothesis that pointed teeth will bite deeper than flat-topped teeth. She makes two 'teeth' from modelling clay, one with a pointed top and one with a flat top. She plans to drop each 'tooth' into a bowl of flour and measure the depth of the hole each one makes. She will drop each tooth so that the top is the first part to go into the flour.

How can she make the test fair?

....

....

....

Respiration

4 Which of the following statements about respiration are true? Tick (✔) **two** boxes.

All living things respire. ☐

In respiration energy is taken in from food. ☐

Respiration is the process of moving air in and out of the body. ☐

Many aquatic animals have gills which take up oxygen dissolved in water. ☐

Movement

5 a What are the structures in the body that allow animals to move?

....

b Give **three** reasons why animals move.

....

....

....

c The heart is made almost entirely of these structures. What do they allow the heart to do?

....

....

d What do these structures do in the stomach?

....

....

Sensitivity

6   Name the **five** sense organs and for each one state the sense with which it is associated.

....

....

....

....

....

Growth and reproduction

7   The table below shows the growth of four female elephants over 20 years.

Elephant	0 years Shoulder height/cm	5 years Shoulder height/cm	10 years Shoulder height/cm	15 years Shoulder height/cm	20 years Shoulder height/cm
Abena	100	148	170	185	190
Bisa	120	165	200	205	210
Chipo	130	180	210	215	218
Doli	110	155	185	200	205

a   Which elephant is the smallest? ....

b   Which elephant is the largest? ....

c   What was the increase in shoulder height of Bisa over the first 15 years of her life?

....

d By how much did each elephant grow in the first 5 years of her life?

....

....

....

....

e How much did each elephant grow between 15 and 20 years?

....

....

....

....

f Describe the pattern shown by the growth of the elephants over 20 years.

....

....

....

....

g Predict the shoulder height of Doli at age 25.

....

....

Excretion

8 a When food and oxygen are used up in the body, waste products are made. Explain why the body must get rid of them.

....

....

....

....

....

....

b Name **two** liquids that the human body excretes.

....

....

c What is the waste product we release in the air we breathe out?

....

Plant life

9 a What signs of life are being shown by the plant on the right?

....

....

....

b When a new plant spreads out over the ground, what **two** signs of life does it display?

....

....

10 Does the colour of light affect the way a plant grows towards it? How could you find out using the equipment shown in the diagram on the right? You will also need to use other materials to help you investigate. You may need to make more sets of equipment like that in the diagram.

box with hole
cut in one end

lid

baffles held in place
with sticky tape

Plan an experiment to investigate the effect of different colours on the plant.

....

....

....

....

....

....

....

....

In your enquiry, what is

a the independent variable(s)? ....

....

b the dependent variable(s)? ....

....

c the control variable(s)? ....

....

Looking for life beyond the Earth

11 John is setting up an experiment to be carried on a space craft which is travelling to another planet. He wants to find out if there are living things in the soil on the other planet.

a What could he add to the soil which might be used by living things that are in it?

....

....

b What could he test for to see if there are any living things in the soil?

....

....

c The space craft lands in just one place on the planet, but will be able to make a few soil tests around it. If the tests do not show signs of life, is the conclusion 'There is no life on the planet' accurate, or is it limited? Explain your answer.

....

....

....

....

2 Identifying species

1 a What do you understand by the word 'species'?

....

b Describe the offspring produced when a male of one species reproduces with the female of another species.

....

c What is a mule?

....

Making observations

2 Observe one of your fingers and make a biological drawing of it.

On your drawing, label the nail and the joints, and give an indication of its size (for example, life-size is $\times 1$ and twice life-size is $\times 2$).

What are the strengths and limitations of the model you have just produced?

....

....

....

....

....

....

....

....

Observing and classifying

3 a Put a tick (✔) in **one** box in each row to distinguish between plants and animals.

Feature	Plant	Animal
cannot make own food		
has cellulose for support		
has chlorophyll		
can move about		

b What does the word 'classify' mean?

....

c Give an example of some things that have been classified by scientists.

....

....

4 Carl Linnaeus classified living things.

a He put living things which were very similar into a group with a special name. What is it?

....

b He made up names of the living things using two languages. What are they?

....

c Why is the work done by Linnaeus important to scientists studying living things in different parts of the world?

....

....

5 A slug has a hydrostatic skeleton. This means that it has water inside its body which acts like a skeleton and gives it support.

    a How could you use a long balloon and a jug of water to make a model of a hydrostatic skeleton in a slug?

    ....

    ....

    b How could you test your model to see if it can supply support?

    ....

    c What are the strengths and limitations of the model?

    ....

    ....

Keys

6 Here are the descriptions of four animals.

    A six legs, blue body   **C** eight legs, brown body

    B eight legs, grey body   **D** six legs, green body

Make a spider key in the box below. Begin by using the words in the box to make your first division.

                                           Animal body

7

A

B

C

D

Use the key below to identify three of the birds in the diagram. In each case write down the number of each statement you used to make the identification. The first bird identification has been done for you, to show how you should set out your answers.

A = 1a, 2b. A is an ibis.

1	a	bird with long beak	*See 2*
	b	bird with short beak	*See 3*
2	a	bird with upturned beak	Avocet
	b	bird with downturned beak	Ibis
3	a	bird with pointed beak	Warbler
	b	bird with blunt beak	Finch

....

....

....

....

8 Make a spider key to identify the four imaginary animals listed below.

A curly shell, six legs and big eyes

B curly shell, eight legs and little eyes

C pointed shell, six legs and little eyes

D pointed shell, eight legs and big eyes

3 | Cells

What is a cell?

1 a When Robert Hooke used a microscope to look at a thin sheet of cork, what did he see?

....

....

b What did he call his findings?

....

....

c Why did he use that name?

....

....

d When Robert Brown examined plant cells with a microscope, he found a structure inside which means 'little nut'. What is the word we use for this structure today?

....

The microscope

2 What equipment do you need to carry out a microscopic examination of a pond weed leaf?

....

....

....

....

....

Basic parts of a cell

3 a Label the parts of this animal cell.

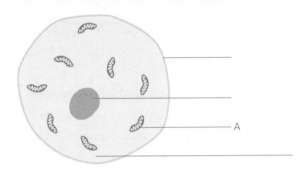

b What is the purpose of the part labelled A?

....

c What else would you find in a typical plant cell?

....

 4 How could you make a model of an animal cell from a cardboard box, a blanket and a small football?

a Draw a picture of the model you would make.

b Label the parts of your model, stating the material and the cell feature it represents.

c How is your model different from a real animal cell? ....

....

....

d Describe the strengths and limitations of your model.

....

....

....

....

 5 Many cells can reproduce themselves. An experiment was set up to find out how cell numbers changed over 4 hours. The graph shows the results.

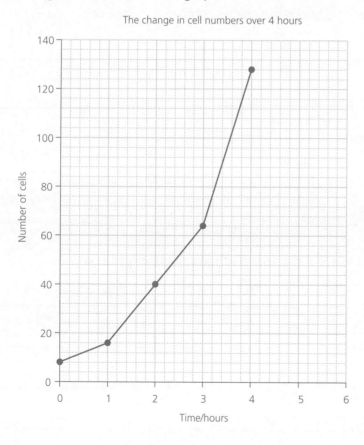

The change in cell numbers over 4 hours

a Make a table in this space and record the data from the graph.

b Identify the pattern revealed by the data.

....

....

c Which result seems to be anomalous as it doesn't fit the pattern?

....

....

d Predict the number of cells after 6 hours.

....

....

e What two experiments could you do to see if the speed of reproduction could be

i increased? ....

....

....

ii decreased? ....

....

Adaptation in cells

6

    a Where precisely would you find palisade cells?

    ....

    b They contain large numbers of a certain cell structure – what structure?

    ....

    c What is the purpose of this structure?

    ....

7 a In the space below, draw a red blood cell.

Red blood cell

    b Describe the red blood cell's shape.

    ....

    c What disappears from inside the cell when it is fully grown?

    ....

    d What is the name of the substance that combines with oxygen when the cell passes through the lungs?

    ....

8

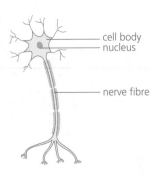

cell body
nucleus

nerve fibre

   a What kind of cell is shown in the diagram above?

....

   b What is an alternative name for this kind of cell?

....

   c What passes along the length of the cell?

....

9

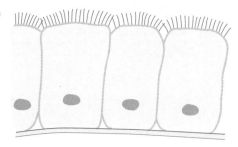

   a Where are these cells found?

....

....

   b What are the structures on the top surface of the cells?

....

....

   c What do these structures do?

....

....

....

10 a What do you understand by the word 'adaptation'?

....

....

  b Give an example of an adaptation in a plant cell and explain what it does.

....

....

....

....

....

....

  c Give an example of an adaptation in an animal cell and explain what it does.

....

....

....

....

Cells, tissues, organs and organisms

11 What is a tissue? Tick (✔) **one** box.

A group of different kinds of cells that do different tasks ☐

A group of different kinds of cells that do a special task ☐

A group of the same kind of cells that do different tasks ☐

A group of the same kind of cells that do a special task ☐

12 What is the difference between an organ system and an organism? Answer by completing these sentences.

a An organ system is ....

....

....

....

b An organism is ....

....

....

....

Microorganisms

13 Use your knowledge of cells to explain the difference between an organism and a microorganism.

....

....

....

....

....

Viruses

14 Why are scientists not certain that viruses are living things? ....

....

15 The diagram below shows a cell infected with viruses.

a What would you use to make a model of the infected cell?

....

....

....

....

....

....

b Describe the strengths and limitations of this model.

....

....

....

....

....

4 Microorganisms

Decomposers

1 How are the bodies of most microorganisms different from the bodies of plants and animals?

....

....

2 a What do decomposers use for food?

....

....

   b How do decomposers in a habitat

     i help the plants that live there?

....

....

....

     ii help the herbivores that live there?

....

....

....

     iii help the predators that live there?

....

....

....

3 a Jane thinks that decomposers in soil will break down fruit faster than they will break down vegetables.

Is this hypothesis testable? Explain your answer.

....

....

....

....

b Mark thinks that decomposers break down food better when the stars are shining.

Is this hypothesis testable? Explain your answer.

....

....

....

....

c Clare thinks that decomposers' actions could be due to the spin of the Earth.

Is this hypothesis testable? Explain your answer.

....

....

....

....

4 Carina has two compost bins. One is in the sun all day and the other is in the shade. She plans an enquiry to test her hypothesis that 'Grass cuttings will break down faster in the shady compost bin than the bin in the sun, because it is cooler in the shade and more decomposers will survive there'.

a Describe a fair test she could make to test her hypothesis.

....

....

....

b When Carina collects her data and presents her evidence, what would you expect her evidence to be if

    i it supported the hypothesis?

    ....

    ....

    ....

    ii it contradicted the hypothesis?

    ....

    ....

    ....

5 In a science laboratory there are two containers of microorganisms. One container has this symbol on it:

a What does the symbol mean?

....

....

The second container has this symbol on it:

b What does this symbol mean?

....

....

c Which microorganisms could damage an ecosystem?

....

....

6 A scientist sets up one solution of yeast and sugar, then puts samples taken from this single solution in test tubes that are kept at different temperatures. She records the height of the froth produced in each of the sample tubes after 20 minutes. Here is a table of her results.

Temperature of sample/°C	Height of froth mm
0	0
20	15
30	40
40	35
50	0

a Plot the data that is shown in this table on the grid below, and draw a line graph.

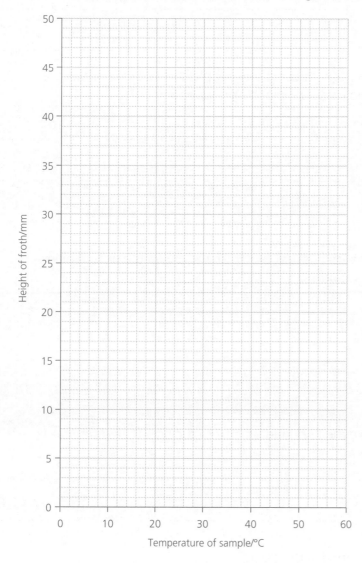

b How did the scientist prepare the samples to make sure that the test was fair?

....

....

c The scientist warmed up the sample at 0 °C and it made a froth. She cooled down the sample at 50 °C but it did **not** produce a froth. Explain the results.

....

....

....

d In the investigation, which is

i the independent variable? ....

ii the dependent variable? ....

7 a When Louis Pasteur began his scientific career, many people believed in an idea called spontaneous generation. What did these people think happened?

....

....

....

b Pasteur made many experiments with flasks of broth like the one in the picture below. He found that the broth did not go bad.

swan-necked
flask

broth did not
go bad

sterile broth
at the start

i What happened when he broke the neck and let air into the broth?

....

....

ii How did Pasteur explain the change in the broth?

....

....

c A process called pasteurisation is named after him. What does it do to microorganisms in liquids like milk?

....

....

....

d Pasteur also did work that supported germ theory.

   i What is germ theory?

....

....

....

   ii How did Pasteur's work and germ theory lead to many other scientists' discoveries?

....

....

....

....

Microorganisms in food chains and food webs

8 Here is a food chain.

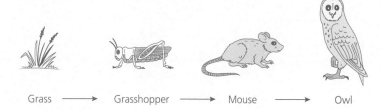

Grass ⟶ Grasshopper ⟶ Mouse ⟶ Owl

Add decomposers to the food chain diagram by writing down the word 'decomposers' and using one or more arrows as you think necessary.

9 Here is a food web.

Write the word 'decomposers' beneath the wildflowers and the grass, and draw arrows to show the path of food to them when the mouse, hawk and wildflowers die.

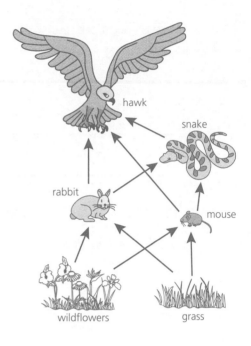

10 Decomposers recycle minerals. What does this statement mean?

....

....

....

....

....

....

11 How can soil scientists help farmers?

....

....

....

....

5 The states of matter

Matter everywhere and The particle theory of matter

1 Which of the following is a property of a gas? Tick (✔) **one** box.

fixed shape ☐   easy to compress ☐

does not flow ☐   volume does not change ☐

2 Which of the following is a property of a solid that makes it different from a liquid?
Tick (✔) **one** box.

fixed shape ☐   easy to compress ☐

does not flow ☐   volume does not change ☐

3 Which of the following is a property of a liquid? Tick (✔) **one** box.

fixed shape ☐   easy to compress ☐

does not flow ☐   volume does not change ☐

 4 a In the box, draw the arrangement of particles in a solid.

b Write a description of the particle arrangement in the solid.

....

....

....

....

5 a In the box, draw the arrangement of particles in a liquid.

b Write a description of the particle arrangement in the liquid.

....

....

....

....

6 a In the box, draw the arrangement of particles in a gas.

b Write a description of the particle arrangement in the gas.

....

....

....

....

7 Rong has a box lid and some marbles. She uses these items to make models of a solid, a liquid and a gas. She turns the box lid upside down so its sides form four walls to hold the marbles in.

In her model of the solid, all the marbles are packed tightly together so they do not move.

a What are the strengths and limitations of this model?

....

....

....

....

....

In her model of the liquid, Rong takes out half the marbles and tips up the lid so that some marbles can roll over the others.

b What are the strengths and limitations of this model?

....

....

....

....

....

In her model of the gas, Rong takes out most of the marbles and moves the lid from side to side so the marbles roll about, bounce off each other and bounce off the walls made by the lid.

c What are the strengths and limitations of this model?

....

....

....

....

....

A vacuum

8 a What is a vacuum?

....

....

....

b Space is often described as a vacuum, but that description is not correct. Explain why.

....

....

....

9 You have a box lid and some marbles. How would you use these items to make a model of

a a complete vacuum?

....

....

....

b a partial vacuum?

....

....

....

10 a What did Aristotle believe about a vacuum?

....

....

b How did Otto von Guericke test Aristotle's belief?

....

The discovery of the elements

1 What kind of substance is an element?

....

....

2 Which of these elements were known before 1669 and were used in ancient civilisations? Put a circle around your answers.

- calcium
- gold
- copper
- chromium
- iron
- tin
- platinum
- lead
- aluminium
- nickel
- zinc
- chlorine
- silicon
- oxygen
- sulfur
- neon
- silver

3 Luigi Galvani discovered that some chemical reactions that made new substances also produced an electric current. Later Alessandro Volta invented a battery that could be used to make an electric current.

Humphrey Davy reasoned that as an electric current is produced when some new substances are made, then perhaps an electric current could be used to break down a substance into its elements.

Davy made a battery, passed an electric current through water and broke it up into the two elements from which it is formed. He then investigated with a substance called potash, using an electric current to see if he could find the elements it was made from. He mixed the potash with a large amount of water but it did not break up. He continued his investigation by heating some solid potash on a platinum spoon and trying again, but it did not work. Next, he added a small amount of water to the potash, just to make it moist, and this time the electric current passed through it, broke it up and revealed that the potash contained a new element, potassium.

a What was the evidence from Galvani's investigation that Davy used in his own?

....

....

b What invention from Volta's work did Davy use in his own work?

....

c In the text above, underline the part that shows Davy's creative thought.

d What are the two elements that join together to produce water?

....

e What do you think was the prediction that drove Davy to investigate potash?

....

....

....

f Scientists have to be persistent in their investigations. Circle **two** sections of text on the previous page to highlight **two** activities that show Davy's persistence in finding an answer to his prediction.

The physical properties of elements

4 Which of these statements about elements and atoms is true? Tick (✔) **one** box.

An atom is made from elements. ☐

Many elements have the same type of atom. ☐

An element is made from one type of atom. ☐

An element is made from many different types of atom. ☐

5 Which of these statements about elements is true? Tick (✔) **one** box.

An element is made from simpler substances. ☐

An element can be split into simpler substances by chemical reactions. ☐

An element is split into simpler substances by physical processes. ☐

An element cannot be split up into simpler substances by chemical reactions. ☐

6 Name **two** elements that are liquid at room temperature and describe their colour.

Element	Colour

7 Name **five** elements that are gases.

....

....

....

....

....

8 Match each element with the colour it produces when heated by drawing a line between them.

Element

| calcium |
| copper |
| magnesium |
| sodium |

Colour

| green |
| golden yellow |
| orange |
| white |

9 Below are the steps in the flame test, but they are in the wrong order. Arrange the steps in the correct order by drawing a line between each step and its true position in the test.

Hold the dipped end of the wire in the Bunsen flame.	1
Set the Bunsen burner on a small flame.	2
Dip the wire in dilute hydrochloric acid to clean it for the next test.	3
Note the colour of the flame on the end of the wire.	4
Dip one end of a clean nichrome wire into the sample solution.	5
Light the Bunsen burner.	6

Chemical symbols

10 Here is a page of alchemists' symbols.

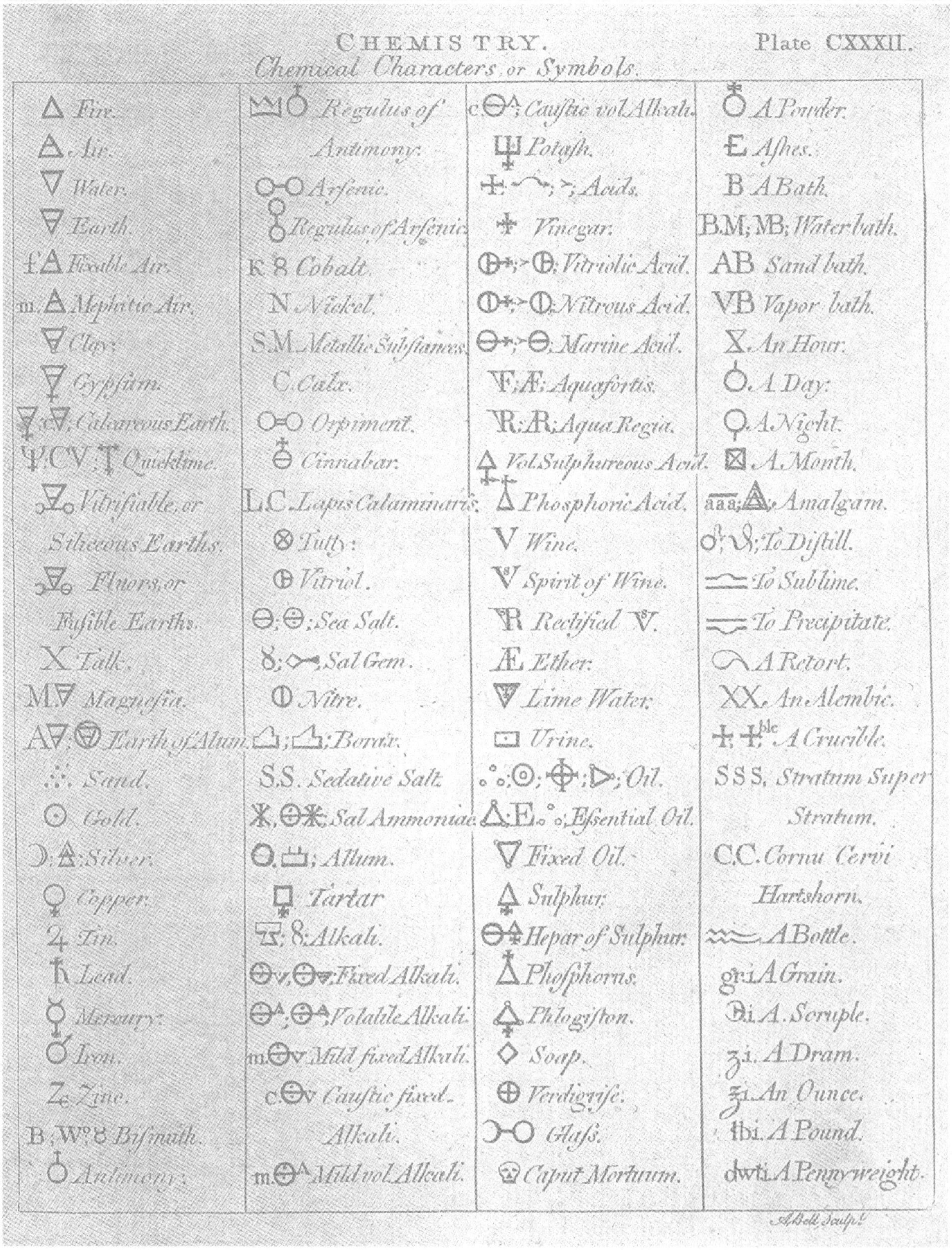

Here are some alchemist's instructions for how to perform an investigation.

Translate the symbols into English and write down the instructions in the spaces below.

Collect ....

....

Mix in ....

for ....

Heat in ....

for ....

Leave for ....

Then put in ....

Collect	⊙ ♀ ☿ ⬚ ✛ ◇
Mix in	┼ble for ⋈
Heat in	B for ⚲
Leave for	⊠
Then put in	〜〜

11

																	VIII
							¹ H hydrogen										² He helium
I	II											III	IV	V	VI	VII	
³ Li lithium	⁴ Be beryllium											⁵ B boron	⁶ C carbon	⁷ N nitrogen	⁸ O oxygen	⁹ F fluorine	¹⁰ Ne neon
¹¹ Na sodium	¹² Mg magnesium											¹³ Al aluminium	¹⁴ Si silicon	¹⁵ P phosphorus	¹⁶ S sulfur	¹⁷ Cl chlorine	¹⁸ Ar argon
¹⁹ K potassium	²⁰ Ca calcium	²¹ Sc scandium	²² Ti titanium	²³ V vanadium	²⁴ Cr chromium	²⁵ Mn manganese	²⁶ Fe iron	²⁷ Co cobalt	²⁸ Ni nickel	²⁹ Cu copper	³⁰ Zn zinc	³¹ Ga gallium	³² Ge germanium	³³ As arsenic	³⁴ Se selenium	³⁵ Br bromine	³⁶ Kr krypton
³⁷ Rb rubidium	³⁸ Sr strontium	³⁹ Y yttrium	⁴⁰ Zr zirconium	⁴¹ Nb niobium	⁴² Mo molybdenum	⁴³ Tc technetium	⁴⁴ Ru ruthenium	⁴⁵ Rh rhodium	⁴⁶ Pd palladium	⁴⁷ Ag silver	⁴⁸ Cd cadmium	⁴⁹ In indium	⁵⁰ Sn tin	⁵¹ Sb antimony	⁵² Te tellurium	⁵³ I iodine	⁵⁴ Xe xenon
⁵⁵ Cs caesium	⁵⁶ Ba barium	⁵⁷ La lanthanum	⁷² Hf hafnium	⁷³ Ta tantalum	⁷⁴ W tungsten	⁷⁵ Re rhenium	⁷⁶ Os osmium	⁷⁷ Ir iridium	⁷⁸ Pt platinum	⁷⁹ Au gold	⁸⁰ Hg mercury	⁸¹ Tl thallium	⁸² Pb lead	⁸³ Bi bismuth	⁸⁴ Po polonium	⁸⁵ At astatine	⁸⁶ Rn radon

a What is shown in the diagram? ....

b How did it get its name?

....

....

c What do the letters stand for in each of the boxes?

....

Elements, compounds and mixtures

Elements, atoms and compounds

1 What is the difference between an element and a compound?

....

....

2 What is the difference between an element and a mixture?

....

....

3 What is the difference between a compound and a mixture?

....

....

4 a Box A shows the arrangement of particles in a pure metal. In box B, draw the arrangement of particles when a second metal is added to make an alloy.

A

B

b Explain why the alloy is harder than the pure metal.

....

....

Mixing elements and making compounds

 5 a Raul has a sample of sulfur and a sample of iron. He observes their colours, tests them with a magnet and mixes them with water.

Record the properties he should have discovered in his investigation.

Property	Iron	Sulfur
Colour		
Test with a magnet		
Mixing with water		

Raul takes some iron and sulfur powder and mixes them together.

b What happens when he brings a magnet near the mixture?

....

c What happens when he stirs the mixture in water?

....

d Raul gathers apparatus to turn the elements into a compound and they are listed below. Describe the use of each piece of apparatus.

Bunsen burner: ....

Crucible: ....

Pipe clay triangle: ....

Tripod: ....

e Describe the compound formed.

....

....

f Name the compound formed.

....

Shamia is making a presentation using models to explain what happened in Raul's experiment with iron and sulfur. She is using balls of yellow modelling clay to represent sulfur atoms and balls of black modelling clay to represent atoms of iron. In iron sulphide, one atom of iron joins with one atom of sulfur.

g How could she set out to show a model of

i a sample of sulfur?

....

....

ii a sample of iron?

....

....

iii a sample of iron sulfide?

....

....

h What are the strengths and limitations of

i the model of sulfur?

....

....

....

....

    ii the model of iron?

....

....

....

....

    iii the model of iron sulfide?

....

....

....

....

6 Carbon dioxide is a compound formed from one atom of carbon and two atoms of oxygen.

    a How could you make a model of this compound? Draw your idea in the box.

    b What are the strengths and limitations of the model?

....

....

....

....

7 Methane is a compound made from one atom of carbon and four atoms of hydrogen.

    a How could you make a model of this compound? Draw your idea in the box.

    b What are the strengths and limitations of the model?

....

....

....

....

....

Different types of mixture

  8 Here are descriptions of three kinds of mixtures. For each one, draw a model in the box and state its strengths and limitations.

    a A foam has gas bubbles trapped in a liquid.

        i Make a model (drawing) in the box.

        ii What are the strengths and limitations of the model?

....

....

....

....

....

b A suspension is made of tiny insoluble particles spread out through a liquid.

   i Make a model (drawing) in the box.

   ii What are the strengths and limitations of the model?

....

....

....

....

....

c A mist is made from tiny water droplets mixed with air.

   i Make a model (drawing) in the box.

   ii What are the strengths and limitations of the model?

....

....

....

....

....

Alloys

9 Which of the following statements about alloys are true? Tick (✔) **two** boxes.

An alloy is a metallic substance made from one metal element. ☐

An alloy is a metallic substance made from a compound. ☐

An alloy is a metallic substance usually made from a mixture of two or more elements which are metals. ☐

Steel is an alloy of iron and the non-metal carbon. ☐

10 Bronze and brass are two alloys.

   a What is bronze made from?

   ....

   b Give an example of a use of bronze.

   ....

   c What is brass made from?

   ....

   d Give an example of a use of brass.

   ....

11 A   B

   a Which diagram, A or B, shows the arrangement of atoms in an alloy?

   ....

   b How can you tell which one is the alloy?

   ....

   c How does the arrangement of atoms make the alloy hard?

   ....

   ....

   ....

8 Physical properties of matter

Metals and non-metals

 1 a Substance A has a high melting and boiling point while substance B has a low melting and boiling point. Identify each substance by putting ticks (✔) in the most appropriate boxes.

Substance	Metal	Non-metal
A		
B		

b Name **two** non-metals found in the air.

1 ....

2 ....

c How is the surface of a metal different from the surface of a non-metal?

....

....

2 Match the five non-metals to their uses by drawing lines between them.

Non-metal	Uses
carbon	car tyres
chlorine	portable water-purifying kits
iodine	barbecue charcoal
phosphorus	keeping swimming pool water clean
sulfur	matches

The properties of materials

3 Some physical properties of materials are listed below.

- rigid
- flexible
- brittle
- absorbent

- transparent
- translucent
- opaque
- electrical conductor

- electrical insulator
- heat conductor
- heat insulator

a Identify the properties of the following materials from the descriptions given.

Material A: It cannot be bent but it snaps if a large force is applied to it. You can see through it.

Properties ....

Material B: You cannot see through it; it takes up water and bends.

Properties ....

Material C: You cannot see through it clearly; it cannot be bent and does not let heat pass along it.

Properties ....

Material D: You cannot see through it; it bends and when placed in a circuit it makes a lamp light up.

Properties ....

b Which material could be a metal?

....

c Which material could be glass?

....

4 Three materials feel soft. You can use a large heavy ball and other items to compare the softness of the materials.

a What hypothesis could you use to test the softness of each of the three materials?

....

....

b Construct a prediction based on the hypothesis.

....

....

....

....

c State your plan to test the softness of the materials.

....

....

....

....

d What steps will you take to produce reliable observations and measurements?

....

....

e How will you find out if the evidence supports or contradicts the hypothesis?

....

....

f What may limit your conclusion?

....

....

....

5 You have four twigs. Each one is from a different species of tree. You also have a number of 10 g masses, a small plastic bucket, two laboratory stools and a metre rule.
Plan an experiment to find out which one is the most rigid and which one is the most flexible. What control variables do you need to think about?

....

....

....

....

....

....

In your experiment, what is

a the independent variable(s)? ....

....

b the dependent variable(s)? ....

....

c From your data, how will you tell which is most flexible and which is most rigid?

....

....

....

....

6 Ahmed carried out an investigation to find out which cloth absorbed the most water. After checking the mass of each material, he dipped each one in water for ten seconds, making sure each piece of material was fully submerged. He allowed the water to drop off the material for 30 seconds then found the mass again. The table shows the data he collected.

Cloth	Mass of cloth/g	Mass of cloth and water/g	Increase in mass due to water/g
A	100	112	
B	100	110	
C	100	116	
D	100	108	

a Did Ahmed make a fair test? Explain your answer.

....

....

b Calculate the increase in mass due to the water and complete the table on the previous page. Use this information to make a bar chart of the increase in mass of the four cloths on the grid below.

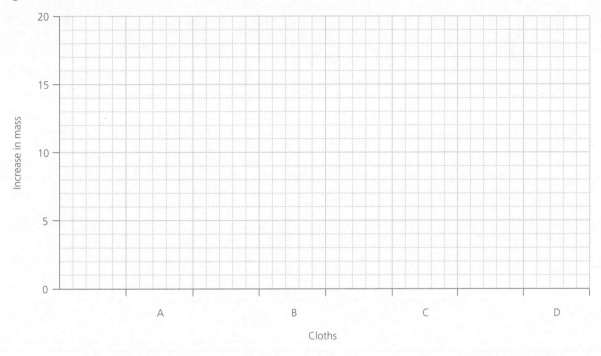

c List the cloths in order of their absorbency, starting with the most absorbent.

....

....

d Could Ahmed's experiment be improved to produce more reliable data?

....

....

7 a How would you use modelling clay to make a model of a sample of an absorbent material and a model of a sample of a non absorbent material? Answer by drawing your models in the appropriate boxes below.

Box A: Absorbent material

Box A: Non-absorbent material

b Explain how the answer to **a** helps to explain the difference between the two materials.

....

....

....

....

....

....

....

Property profiles

8 A cooking pan is made of metal and has a wooden handle.

a What properties of the metal make it useful for making the cooking pan?

....

....

b What properties of the wood make it useful for making the handle?

....

....

....

9 Three rods are dipped in water at 35 °C. The other end of each rod is attached to a temperature sensor (an electronic device that measures the temperature of its surroundings) and the temperature is recorded for 5 minutes. The data is shown in the table below.

Rod	0 min	1 min	2 min	3 min	4 min	5 min
A	20 °C	21 °C	22 °C	23 °C	24 °C	25 °C
B	20 °C	20 °C	20 °C	20 °C	20 °C	20 °C
C	20 °C	22 °C	24 °C	26 °C	28 °C	30 °C

a Which rod is made of a heat-insulating material?

....

b Which rod is made of a better heat-conducting material?

....

c On the grid, plot the change in temperature of the better heat-conducting material.

d Predict or estimate how long it would take the end of this rod to reach the same temperature as the water (35 °C)? Use your graph to find out.

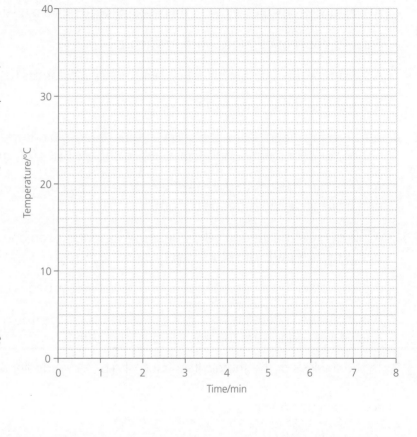

....

....

9 Chemical reactions

When a candle burns

1 Which of the following statements about chemical reactions is true? Tick (✔) **one** box.

A substance combines with another substance to make a mixture. ☐

A substance changes state when it is heated. ☐

A substance takes part in an activity with other substances to make new substances. ☐

A substance changes when it is cooled. ☐

2 Jane is using green beans and yellow beans to make a particle model of a chemical reaction. She makes four models. Which one should she use? Tick (✔) **one** box.

A pile of green beans and a pile of yellow beans on one side and a pile of green beans and a pile of yellow beans on the other. ☐

A pile of green beans and yellow beans stuck together on one side and a pile of green beans and yellow beans stuck together on the other. ☐

A pile of green beans on one side and a pile of yellow beans on the other. ☐

A pile of green beans and a pile of yellow beans on one side and a pile of green beans and yellow beans stuck together on the other side. ☐

3 In the analysis of the chemical reaction of a burning candle, what are the following used for?

a The suction pump ....

....

b The cobalt chloride paper ....

....

c The limewater ....

....

4 Two birthday candles the same size and mass, A and B, are lit and placed in dishes on a bench. A large beaker is upturned and placed over candle A.

After a few minutes, candle A stops burning but candle B keeps burning.

a In which candle has the wax stopped taking part in a chemical reaction?

....

b Explain why A stopped burning but B continued burning.

....

....

....

c What would you predict to happen to candle A if it was relit and a smaller beaker was placed over it?

....

....

....

d Make a plan to find out if the size of a beaker over a candle affects the length of time it burns. What control variables do you need to think about?

....

....

....

....

....

....

In your experiment, what is

i the independent variable(s)? ....

....

ii the dependent variable(s)? ....

....

The chemical reaction when calcium hydroxide (limewater) turns milky

5 a What is the calcium compound in limewater?

....

b What calcium compound makes limewater white?

....

6 A flask is set up with a chemical that can produce carbon dioxide gas in a chemical reaction. There is a delivery tube attached to the flask which dips into a test tube next to it.

a What could be put in the test tube to see if carbon dioxide is produced when a second chemical is added to the first?

....

b What would you expect to see in the test tube if carbon dioxide was produced in the chemical reaction?

....

....

c How could you use the apparatus and the contents of the test tube to see if one pair of chemicals produced carbon dioxide faster than another pair?

....

....

....

....

....

....

....

....

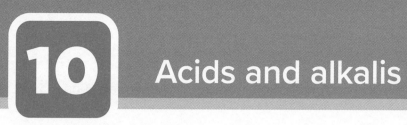

Acids and alkalis

Early acids and alkalis

1 The names 'acid' and 'alkali' come from Latin and Arabic words respectively. What did these words mean? Select one meaning for each of the words from this list.

- sweet   – sour   – bitter
- the plants   – the soil   – the ashes

Acid ....

Alkali ....

Detecting acids and alkalis

2 Who discovered a way of detecting acids and alkalis? Tick (✔) **one** box.

Democritus ☐

Gerber ☐

Boyle ☐

Sorensen ☐

Using indicators

3 a On the scale below mark an X where the pH is neutral.

b Circle a group of **three** divisions that indicate a strong acid.

c Circle a group of **four** divisions that indicate a weak alkali.

| 0 | 1 | 2 | 3 | 4 | 5 | 6 | 7 | 8 | 9 | 10 | 11 | 12 | 13 | 14 |

4 What colour is litmus in a neutral solution?

....

Neutralisation

 5 a What happens to the pH when an acid and an alkali are mixed during a neutralisation reaction?

....

....

b If Universal indicator was added to a solution of an acid, an alkali and a mixture of them, what colour would you expect to be in

  i the acid?

....

  ii the alkali?

....

  iii the products of the reaction?

....

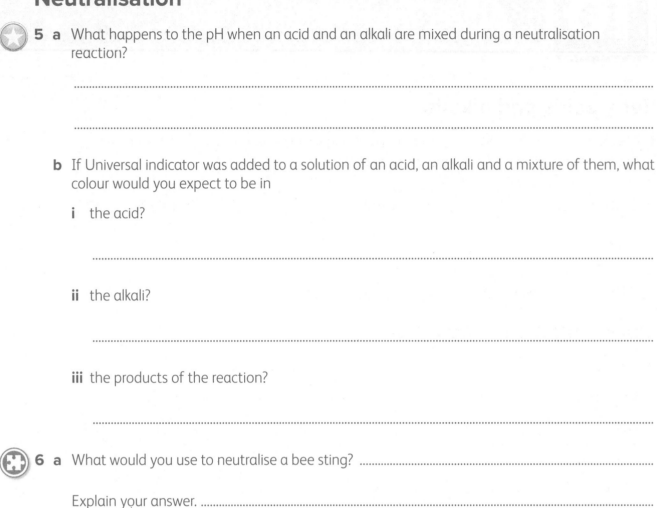 **6 a** What would you use to neutralise a bee sting? ....

Explain your answer. ....

....

b What would you use to neutralise a wasp sting? ....

Explain your answer. ....

....

7 A sample of vinegar (acetic acid) has a pH of 3. Water has a pH of 7.

   a When water is added to vinegar to make a mixture, do you think the pH of the mixture will go down (become a lower number like 2) or go up (become a higher number like 5)?

   ....

   ....

An acid with a lower pH, such as 2, has stronger acidic properties than an acid with a higher pH, such as 4. When an acid is added to sodium bicarbonate, a reaction takes place which produces lots of bubbles of carbon dioxide.

   b You have a sample of vinegar and a sample of vinegar which has been mixed with water. Which sample do you predict will make the most bubbles when added to sodium bicarbonate?

   ....

   ....

   c Plan an experiment to test your prediction in part **b**. In your experiment which is the variable

   i that you change?

   ....

   ii that you measure?

   ....

   iii that remains the same?

   ....

8 a What causes indigestion? ....

   ....

   b Why is sodium bicarbonate used in tablets to cure indigestion?

   ....

   ....

9 a Why does baking powder need water to make it produce a chemical reaction?

....

....

b What is the useful product of this reaction and how does it affect a cake?

....

....

10 a What kind of substance is a precipitate? Tick (✔) **one** box.

A solid ☐

A liquid ☐

A gas ☐

b Is a precipitate soluble or insoluble in the liquid around it?

....

c If a chemical reaction takes place in a clear liquid which produces a precipitate, how does the appearance of the liquid change?

....

....

d What is the name of the compound that forms in limewater which forms a precipitate when carbon dioxide is bubbled through it?

....

11 What can you conclude from observing the mixing of two clear liquids and seeing a precipitate form?

....

11 Measurement

Length, mass and time

 1 James says he has a hypothesis about the way he kicks a football. He says he is a better kicker with his right foot because he is right-handed.

    a Why is this hypothesis untestable?

    ....

    ....

    b How could you make his hypothesis into a testable one?

    ....

    ....

    c Write down a plan to test your hypothesis. In your plan write down:

        – the equipment and location you will need
        – the measurements you will take
        – the number of measurements you will take
        – the safety precautions you will take
        – in which ways the test could be unfair
        – the ways in which you will try to make sure the test is fair.

    ....

    ....

    ....

    ....

    ....

    ....

    ....

    ....

d How will the evidence you collect support or contradict your hypothesis?

....

....

....

....

e What are the limitations of your conclusion?

....

....

....

....

....

2 Alisha is using a force meter to find out the force needed to pull a mass down a slope when the slope is set to different heights.

Here is a table of her measurements.

Height of one end of slope (cm)	Force needed to start movement down slope (N)
5	30
10	25
15	23
20	15
25	10

a In her enquiry

i what is the dependent variable?

....

....

ii what is the independent variable?

....

....

b Produce a line graph of the results by placing the height of the slope on the X axis and the force needed to start movement on the Y axis.

c Which measurement is an anomalous result?

....

....

....

....

d Is there a pattern or trend in the results? Explain your answer.

....

....

....

....

3 Ben, Oliver and Max are growing bean plants to see how fast they grow. Ben and Max are very competitive. Here are the height measurements of the plants grown in identical conditions.

Day	Ben's plant height (cm)	Oliver's plant height (cm)	Max's plant height (cm)
1	5	5	5
3	9	8	9
5	13	11	14
7	19	16	19.5
10	24	22	24.5

a Shazia is trying to find out about how fast beans grow. Is the table produced by Ben, Oliver and Max relevant to her enquiry? Explain your answer.

....

....

....

b Shazia suspects some of the data may be biased. Why do you think she has these suspicions?

....

....

c What must she do to check the data in the table?

....

....

....

Accuracy of measurements

 4 Bamboo has the fastest growing plant shoots. Growth can be seen to have taken place after a few hours.

Joel and Larisa want to find out if a young bamboo shoot grows faster or slower than an older bamboo shoot.

Joel sets up two plants in the same conditions and measures their height in centimetres and repeats the measurements every day at the same time.

Larisa sets up the two plants in the same conditions and measures their height in centimetres and millimetres early in the day, then repeats the measurements at four-hour intervals for the rest of the day until dark.

a In their enquiries

   i what is the dependent variable?

   ....

   ii what is the independent variable?

   ....

b Are both students making accurate and precise measurements? Explain your answer.

   ....

   ....

   ....

   ....

c Why are accurate and precise measurements important?

   ....

   ....

   ....

   ....

Heat and temperature

5 Han wants to compare the heat produced by two fuels. He sets up two barbecues. In one he places Brand A charcoal and in the other he places Brand B charcoal. He puts water and a thermometer in two pans and places them on the barbeques, then lights the fuels and measures the temperature of the water every 5 minutes, for half an hour.

a What must Han do to make the test fair?

....

....

....

Han recorded his results in a table.

Time/min	Temperature of water heated by Brand A charcoal/°C	Temperature of water heated by Brand B/°C
0	30	30
5	70	40
10	92	60
15	99	90
20	80	99
25	70	92
30	50	90

b Present the data in the table as two graphs on the grids below.

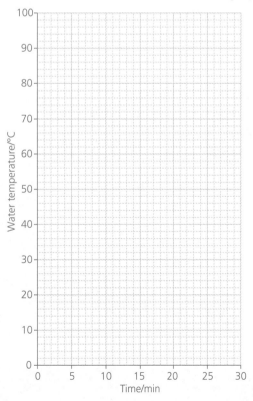

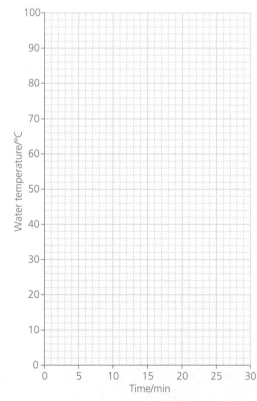

c Describe how the temperature of the water changes with each type of fuel.

   i   Brand A charcoal ....

   ....

   ii Brand B charcoal ....

   ....

d Compare how the fuels release their energy.

   ....

   ....

Stores of energy

1 Clare stands at the side of the swimming pool ready to run onto the diving board.

    a What stored energy is in her body? ....

    Clare runs to the end of the diving board and jumps up.

    b At the top of her jump, what extra stored energy does she possess?

    ....

    Clare lands on the end of the diving board and bends it down.

    c What stored energy does the diving board have now?

    ....

2 Fayola has made a catapult out of an elastic band. She has investigated the stored energy in it in the following way. She placed a cotton wool ball against the band, pulled back the elastic band by 1 cm, then let go and measured the distance the ball travelled. She repeated this process by pulling back the band to different distances and recording her results.

Here is the table of data from her investigation.

Distance band pulled back/cm	Distance ball travelled/cm
1	2
2	4
3	5
4	17
5	29
6	0

a Present the data in the table as a line graph on the grid below.

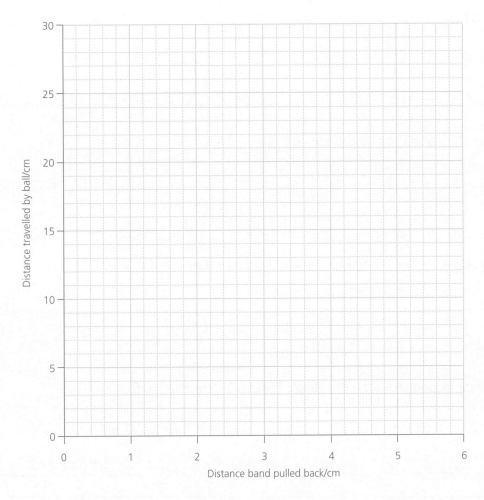

b What pattern does the graph seem to show between the distance pulled back and the distance the ball travelled?

....

....

....

....

c What pattern does the graph seem to show between the amount of stored energy in the band and the distance the ball travelled? Explain your answer.

....

....

....

d One of the results is anomalous.

    i Which one is it?

....

....

    ii Suggest how this result may have been caused.

....

....

e How could Fayola make sure the results show the pattern they seem to show?

....

....

f What do you think happened when the elastic band was stretched to 6 cm?

....

....

3 How can you tell if something has kinetic energy? ....

....

4 State two energy changes that take place when a firework explodes.

    a ....

    b ....

5 Food is a store of chemical energy. When you use this energy in your body to run, one form of energy is useful to you and one is less useful to you. State them.

Useful form of energy ....

Less useful form of energy ....

Energy transfers

6 How could you use the following items to demonstrate strain energy, kinetic energy and sound energy: a catapult, a dried bean, a metal tray and a wall? Below, you may make a diagram of the arrangement of the equipment and explain what you would do.

....

....

....

....

....

7 Materials A and B are dry and can be easily set alight by a flame. How could you compare the chemical energy in them using a weighing scale, scissors, tripod, gauze, a beaker, a measuring cylinder, a thermometer, a clock and water?

Plan an experiment to answer this question. What control variables do you need to think about?

....

....

....

....

....

....

In your experiment, what is

a the independent variable(s)? ....

....

b the dependent variable(s)? ....

....

c Explain how you can compare the chemical energy in the two materials.

....

....

....

....

....

....

....

....

8 The idea behind a perpetual motion machine is that all its energy can be used to make part of it move forever. Explain why the machine cannot work.

....

....

....

....

Wasted energy or energy we cannot use

9 What does dissipation of energy mean? Tick (✔) **one** box.

The gathering together of different forms of energy into one form of energy. ☐

The spreading out of one from of energy. ☐

The spreading out of energy into different forms. ☐

The gathering together of energy that has been spread out. ☐

10 What happens to energy when it dissipates? Tick (✔) **one** box.

It becomes more useful. ☐

It becomes less useful. ☐

Its usefulness does not change. ☐

It becomes completely useless. ☐

13 | Sound

From vibration to sound wave

1 What is a vibration? ....

  ....

2 Name **ten** sources of sound. ....

  ....

  ....

  ....

  ....

3 What happens when you place one end of a ruler over the edge of a table and make it vibrate, then move the end further over the edge and then back again? Tick (✔) **two** boxes.

The sound goes higher as the vibrating length increases. ☐

The sound goes lower as the vibrating length increases. ☐

The sound goes higher as the vibrating length decreases. ☐

The sound goes lower as the vibrating length decreases. ☐

4 Air is made of particles.

  a In Box A, draw how the particles are spread out in the air when there is no sound.

  Box A

b In Box B, draw how the particles are arranged when a vibrating object pushes on them.

Box B

c In Box C, draw how the particles are arranged when the vibrating object moves away from them.

Box C

5 Why do you think the movement of sound is described as a sound wave?

6 **a** How could you use a ruler as a vibrating object and a table tennis ball on a string to show how a sound is generated in air? You may use a diagram in your answer.

b What does the table tennis ball represent?

....

c What are the strengths and limitations of your model?

....

....

....

7 The following statements describe the events that happen as an animal comes near a snake, but they are in the wrong order. Arrange them correctly by writing the letter of each statement in the order in which it occurs.

A The snake's ears receive the vibrations.

B A foot approaches the ground.

C The vibrations pass across the ground.

D The snake prepares to attack.

   E The lower jawbone of the snake starts to vibrate.

   F A foot touches the ground and makes it vibrate.

....

8 If a vibrating object is put in a vacuum, can you hear it? Explain your answer.

....

....

....

Reflection of sound

9 What do you understand by the term 'reflection of sound'?

....

....

Echoes

10 Jane stands just over 17 metres away from a high wall and claps her hands. She hears two sounds.

   a What is the second sound?

....

   b Which of these answers is correct? Tick (✔) **one** box.

     Jane could hear this second type of sound because the ear can only hear two separate sounds if they reach the ear

| together | ☐ | more than a tenth of a second apart | ☐ |
| less than a tenth of a second apart | ☐ | less than a hundredth of a second apart. | ☐ |

   c In a thunderstorm

     i what is the source of the sound of thunder?

....

     ii where is the thunder reflected to make a roll of thunder?

....

14 Electricity

Introducing the electron

1 a Ahmed has made a model of electricity flowing round a circuit with a water pump, pipes connected in a loop to the pump and the pipes full of water. When he switches on the water pump, the water is pushed round the pipes in the loop.

In Ahmed's model what do the following represent?

i The water pump:

....

ii The pipes:

....

iii The water:

....

b Lily has made a dam in a stream. She says she can make it like a model of a circuit with electricity flowing if she unblocks the dam and lets the water go down the stream.

In Lily's model what do the following parts represent?

i The water in the dam:

....

....

ii The water:

....

....

iii The stream:

....

....

c Describe the strengths and limitations of

   i Ahmed's model

   ....

   ....

   ....

   ....

   ii Lily's model.

   ....

   ....

   ....

   ....

2 What is the difference between an electrical conductor and an electrical insulator?

An electrical conductor is ....

An electrical insulator is ....

Simple circuits

3 Naveen sets up the circuit shown in the diagram.

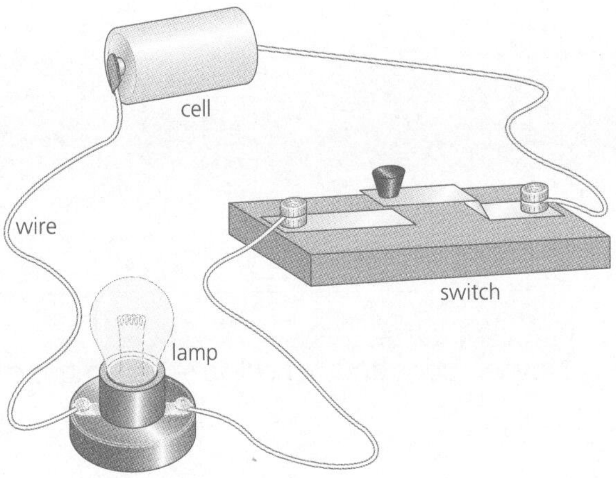

   a In the circuit, you can see the flow of electricity is blocked from moving.

      i Which component is stopping the flow of electricity?

      ....

      ii Why is the component stopping the flow of electricity?

      ....

cell

wire

switch

lamp

b When electricity flows through the circuit, the lamp lights up and Naveen notes its brightness. What happens to the brightness of the lamp when he

    i adds a cell to the circuit?

    ....

    ii takes out the extra cell and adds an extra lamp in series with the first?

    ....

4 Here are the symbols for use in an electrical circuit.

a Look at the picture of Naveen's circuit in question 4 and make a circuit diagram using some of the symbols shown.

Naveen adds another cell in series to the circuit.

b Draw the circuit diagram for this new circuit.

5 The connection of the wire to the lamp breaks. Aruni suggests putting a piece of aluminium foil across the gap. Naveen disagrees and says a piece of wood is all that is needed.

   a Who is correct? ....

   b Explain your answer. ....

     ....

     ....

     ....

6 Sam thinks it is possible to compare the currents flowing in different circuits with different numbers of cells by using a light meter.
Write a plan for him to try, using three cells, wires, a switch, a lamp, a light meter app on a cell phone and a ruler. What control variables do you need to think about?

....

....

....

....

....

In your experiment, what is

a the independent variable(s)? ....

....

b the dependent variable(s)? ....

....

c Advise Sam on what to look for in the results if his idea is correct.

....

....

....

Amperes

7 a What does an ammeter measure? ....

b An ammeter has its positive terminal marked in red. When adding an ammeter to a circuit, which instruction should you follow? Tick (✔) the correct box.

Connect the red terminal to the positive terminal of the cell. ☐

Connect the red terminal to the negative terminal of the cell. ☐

Connect the red terminal to either terminal of the cell. ☐

Connect the red terminal to a lamp. ☐

8 Ochi sets up the circuit shown in the diagram. She connects the ammeter at point A and then at point B.

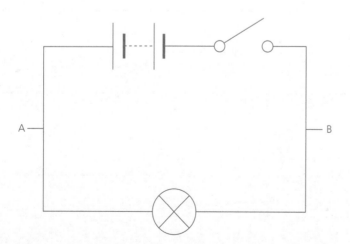

What does she find? Tick (✔) the correct box.

The reading at A is lower than at B. ☐

The reading at A is higher than at B. ☐

The reading at A is the same as at B. ☐

9 a Compare the components in circuit A and circuit B shown below, and make some notes about each.

....

....

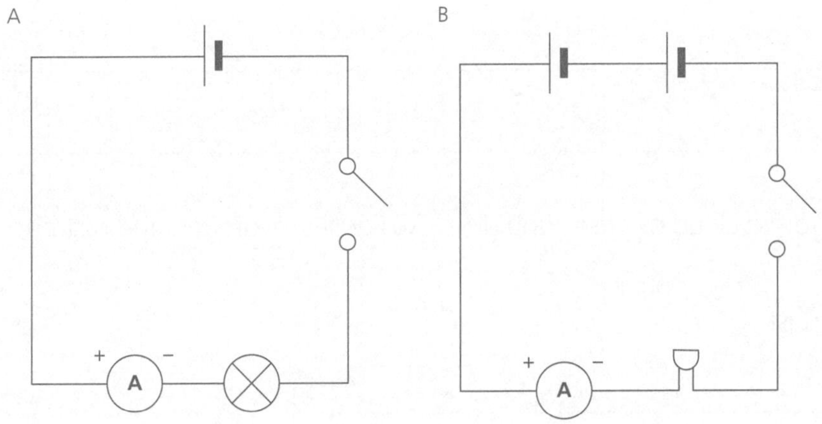

b How do you predict the sound of the buzzer would change if you placed it in circuit A instead of the lamp?

....

....

c How could you explain this change using the readings on the ammeters in the two circuits?

....

....

15 The Earth in space

Gravity and planet formation

1 What is gravity?

....

....

2 A planet can attract things to it by its gravitational force, but a pebble cannot do this. Why?

....

....

3 The solar system formed from a cloud of gases and dust. Two gases, hydrogen and helium, formed the Sun but

    a what happened to the dust?

....

....

....

    b was gravity important in planet formation?

....

....

....

4 How does gravitational force act on objects in the universe?

....

....

....

5 a What is air resistance?

....

....

....

b Does air resistance affect planets moving through space? Explain your answer.

....

....

....

6 How could you use your hand, a piece of string, some sticky tape and a table tennis ball to model the action of a planet going around a star?

a In your model what is the

i planet? ....

ii star? ....

b How will you model the action showing how gravity makes a planet stay in orbit?

....

....

....

....

c How would you demonstrate what would happen if the force of gravity between the planet and star was suddenly to disappear? What would be the result?

....

....

....

....

d What are the strengths and limitations of your model?

....

....

....

The formation of the solar system

7 a What did people first think was at the centre of the solar system?

....

....

b Nicholas Copernicus produced a model of the solar system. How was it different from people's first ideas about the solar system?

....

....

....

c How did Johannes Kepler modify the Copernicus model?

....

....

....

d Who showed that gravity was important in the movement of the planets around the Sun?

....

The Moon, the Sun and the tides

8 The changing shapes of the illuminated surface of the Moon are known as the phases of the Moon. They always occur in the same order.

The phases of the Moon shown here are in the wrong order.

A   B   C   D   E

 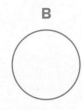   

a Write down the letters of the phases in the order they appear in the sky, starting with the phase straight after the new moon.

....

b What can you see if the Sun, Moon and Earth are in a line?

....

c When the Moon is overhead on the coast, how does its gravitational force affect the tide on the shore?

....

....

9 Which statements are correct? Tick (✔) **two** boxes.

A neap tide occurs when the Sun and Moon are in line.   ☐

A spring tide occurs when the Sun and Moon are in line.   ☐

A neap tide occurs when the Moon is furthest out of line with the Sun.   ☐

A spring tide occurs when the Moon is furthest out of line with the Sun.   ☐

10 In the spaces below, draw and label the Earth, Sun and Moon to show their positions when

    a an eclipse of the Sun occurs.

    b an eclipse of the Moon occurs.

The structure of the Earth

1 There are two layers that make up the solid outer layer of rock on the Earth. What are these two layers called?

....

....

2 What is a tectonic plate?

....

....

The tectonic plates

3 Bo makes a model of tectonic plates and the mantle by floating flat biscuits on a dish of syrup.

   a Which part of the model represents the tectonic plates?

....

....

   b Which part of the model represents the mantle?

....

....

   c What are the strengths and limitations of the model?

....

....

....

....

Bo's teacher places the syrup in a heat-proof dish and floats the biscuits again. She places a candle safely under one end of the dish, and its heat makes the syrup move.

d What might happen to the biscuits when the syrup moves?

....

e What are the strengths and limitations of this model?

....

....

....

....

4 a At what type of boundary does a fold mountain occur?

....

b A fold mountain forms when layers of rock buckle. What do they do in this process to make a mountain?

....

5 An earthquake occurs when stresses in rocks make them crack.

a When can tectonic plates cause an earthquake?

....

....

b Where on a tectonic plate do earthquakes occur? Tick (✔) **one** box.

at the boundaries ☐

in the middle ☐

all over the plate. ☐

6 Is there a link between where you find volcanoes and the tectonic plates?

....

....

....

....

The Earth's atmosphere

7 a What is the composition of nitrogen and oxygen in clean, dry air? Tick (✔) **one** box.

Nitrogen 80 %, oxygen 20 % ☐

Nitrogen 75 %, oxygen 24 % ☐

Nitrogen 78 %, oxygen 21 % ☐

Nitrogen 79 %, oxygen 20 % ☐

b What is the gas produced when living things respire, that is present in small amounts in the atmosphere?

....

c Gases in the atmosphere can be produced by natural emissions. What does this mean?

....

....

....

....

d Name two gases produced by human activities that are causing air pollution.

....

....

Water in the atmosphere and the water cycle

 8 In the space provided, make a drawing of the water cycle and label the following:

A where most evaporation takes place

B where condensation occurs in the sky

C where precipitation occurs

D where run-off occurs

 9 You are provided with a slab of tarmac road surface in a tray, a piece of turf (a patch of grass with roots and soil still attached) in a tray, a tray of pebbles and a tray of soil. They all have the same surface area. You also have a watering can, a large bowl much larger than three trays, a measuring cylinder, a brick, a ruler and a clock.

a Plan an experiment using this equipment to compare the run-off rate of water on the four surfaces. You can measure the run off rate by timing how long it takes for the water to run over the surface of the tarmac, turf, pebbles and soil.

....

....

....

....

....

b Which surface do you predict will have the fastest run-off?

....